FACULTY GUIDE

for use with

SCIENTIFIC
AMERICAN
FRONTIERS

Video Collection for Introductory Psychology
Second Edition

Richard O. Straub
University of Michigan-Dearborn

WORTH PUBLISHERS

Faculty Guide for The Scientific American Frontiers Video Collection for Introductory Psychology, Second Edition by Richard O. Straub

Printed in the United States of America

ISBN 10: 1-57259-903-0
ISBN 13: 978-1-57259-903-1

Seventh printing

Worth Publishers
41 Madison Avenue
New York, NY 10010
www.worthpublishers.com

Scientific American Frontiers
Introductory Psychology Video Collection, Second Edition

Segment	Title
1	Tackling a Killer Disease
2	Water, Water Everywhere
3	Aliens Have Landed?
4	Return to the Wild
5	Mind Reading
6	Image-Guided Surgery
7	Severed Corpus Callosum
8	Old Brain, New Tricks
9	Smart Glasses
10	Lights, Camera, Magic!
11	Cockpit Confusion
12	Tasters and Supertasters
13	Catching Catnaps
14	What's in a Dream?
15	Can You Beat Jet Lag?
16	Remembering What Matters
17	True or False
18	Talkin' Babies
19	If Only They Could Talk!
20	Who Needs Words, Anyway?
21	Born to Talk
22	Bypass Genes
23	Bringing Up Monkey
24	Baby Body Sense
25	The Magic Years
26	A Change of Mind
27	Teaching the Computer to Think
28	Superpower Ping-Pong
29	The Power of Persuasion
30	Virtual Fear
31	Arachnophobia
32	Cop Psychiatrists
33	Sports Imports

Segment 1

Tackling a Killer Disease (8:35)

In the highly regulated world of medicine, it can take ten years or more for a new treatment or drug to be approved. The first step is testing with live animals. If all goes well, the next step is a clinical trial with a small sample of human subjects. One such subject is 10-year-old Justin, whose struggle to survive Duchenne's Muscular Dystrophy is the subject of this heartwarming segment.

Duchenne's is a degenerative, sex-linked disease caused by an abnormal gene. There is no cure, and Justin's family watches sadly as Justin grows weaker. As his muscles weaken, Justin will no longer be able to walk. When his heart and lungs give out, he will die.

In response to normal activity, even healthy muscle cells in healthy humans develop small tears. Myoblasts come to the rescue, fusing with torn cells and producing the protein *dystrophin,* which helps the muscle repair itself. In Justin's form of MD, the cells don't produce dystrophin. When muscle tears occur, there is no healing, and the muscle eventually fails.

Although children like Justin rarely reach their 20s, there is hope that a new experimental treatment will slow, and possibly reverse, the degenerative process. The treatment involves transplanting healthy muscle cells to Justin's muscles to begin producing dystrophin. One of Justin's legs will receive the independent variable—treatment with healthy myoblasts. The other leg will serve as a control and receive injections of an inactive solution. This is a double-blind experiment: not even Justin's doctor knows which leg is receiving the actual treatment.

It will be six months before researchers know if the treatment is working. Even then, the approval process will be far from complete: a full-scale clinical trial with several hundred subjects, and several more years of research to examine possible side effects of the treatment, will be necessary.

Because this segment touches on several topics related to the nature of experimentation, it is an excellent video to use at the beginning of your course.

Topics for Discussion

- the nature of experimentation
- cause-and-effect relationships
- independent and dependent variables
- experimental and control conditions
- placebo
- the double-blind procedure

Discussion Questions

1. What disease is Justin suffering from? How did he acquire this disease? What are the symptoms and mechanisms of the disease? What experimental treatment for this disease is described in the video segment? How is it supposed to work?
2. What type of research study is Justin participating in? What features of the study distinguish it from other research methods that are described in the textbook? Why did the researchers choose this particular method?
3. What are the basic elements of the study that Justin is participating in? What is the independent variable? the dependent variable? What is the experimental condition? the control condition?
4. Assuming that Justin's clinical trial is successful, what is the next step in the approval process for this new treatment? Explain how you would determine whether the results of Justin's chemical trial could be generalized to others suffering from the same disorder.

Segment 2

Water, Water Everywhere (12:20)

Paul Sevigne is dowsing for water in the mountains of Vermont. A nearby ski resort has asked him to find a site for drilling a new well. Like most dowsers, Paul says his mind can directly sense the water, hundreds of feet below. "I'm concentrating on over 25 gallons a minute," he says, "good drinking water." Coming up on a promising site, the rod swoops sharply downward. Mentally asking the dowsing rod about the waters depth and flow rate, Paul confidently states "342 feet deep, 26 gallons per minute." But a little farther on Paul finds an even better site: water at 465 feet, flowing at 48 gallons per minute. This is a better yield and not as deep.

With two promising well sites in hand, Ken Bannister, who works with Paul, uses conventional instruments to locate water, including a device that measures the earth's magnetic field. Small variations can mean fractures in the bedrock, which is where water might be. Ken's not surprised that the two well sites Paul found yield just such magnetic variations. After several years of just such experiences, Ken himself became a believer in dowsing.

In Eugene, Oregon, dowser Jay Todd has been dowsing for water, oil, and lost objects for 20 years. Jay has agreed to allow University of Oregon psychologist Ray Hyman to test scientifically his ability to locate a metal target hidden beneath one of a dozen or so plastic containers scattered across a large field. The procedure begins when another member of the research team uses a calculator to generate a random number electronically. This number indicates under which container the metal pipe, or "receiver," will be hidden. To place the target, another member of the team goes through an elaborate series of fake placements to ensure that Jay is given no clues to the target's location, such as trampled grass. The test is "double-blind," since neither Jay nor any of the people around him during the test know the location of the target.

During the first of four trials, Jay gets a possible response at two locations, neither of which is the correct one. During the second trial Jay asks for a different receiver, this time made of lead, but this doesn't improve his accuracy. For trials three and four, Jay changes dowsing rods. Clearly frustrated, now he gets no response at all.

After years of studying dowsers, Ray Hyman is convinced that there's nothing magical about the ability of dowsers like Jay Todd. He believers that subtle muscle movements, which even the dowsers are unaware of, cause the dowsing rod to move in response to their beliefs and expectations. He demonstrates this explanation by asking host Alan Alda to hold a small weight, suspended first over the palm of his hand, and then over the palm of a female student. Planting the suggestion that over a male hand the weight will swing back and forth and over a female hand it will move in a circle, the weight does exactly that. "I'm not doing anything," exclaims an amazed Alda. "I'm trying to hold it still!"

Applying the results to dowsing, Hyman explains why most dowsers believe their alleged ability is an example of extrasensory perception, or ESP: "Without realizing it they are making this move fit what we think it's expected to. Under the right conditions they think there's an outside force."

Back in Vermont, well driller Dave Parker is not surprised. From his perspective he's been disproving dowsing for years. "Probably 99 percent of the time you're going to hit water, no matter where you drill. It's just a question of how deep you've got to go to get it." At 370 feet there's a trickle of about a gallon per minute. But the dowsers have promised 48 gallons a minute at 465 feet, so they press on. But at 500 feet the flow is no better, and at 600 feet the ski resort calls it off, confirming what Dave thinks of dowsing: "Personally, I have no belief in it. I've proved them wrong too many times."

This segment provides an effective case study for your discussion of the scientific method, especially as it pertains to the importance of experimental control in research. It also touches on subject and experimenter expectancy and several issues pertaining to extrasensory perception.

Topics for Discussion

- the scientific method
- the concept of control
- the double-blind procedure
- extrasensory perception
- subject and experimenter expectancy
- placebo

Discussion Questions

1. What type of research study was used to test Jay Todd's dowsing ability? Why did the researchers choose this particular method?
2. What are the basic elements of the double-blind procedure? Who is "blind," and why? What are some typical scientific uses of this method?
3. How would you set up a double-blind test to evaluate telepathy or another form of ESP? Alternatively, how would you use the double-blind procedure to evaluate a new method of psychological therapy?
4. How do dowsers like Jay Todd explain their alleged ability? What explanation of dowsing does psychologist Ray Hyman offer?

Segment 3

Aliens Have Landed? (11:25)

This captivating segment opens with a black-and-white film of what appears to be an autopsy of an alien space traveler. But as the two "pathologists" remove their surgical gowns, the truth is revealed. It's Alan Alda and a Hollywood special effects expert, working on an incredibly believable movie dummy!

The bogus segment was designed to simulate the television documentary, "Alien Autopsy," which allegedly contained the first evidence of an earthly visit of beings from outer space. This film—a segment of which is also shown here—was supposedly shot by a military photographer following the 1947 crash of a spaceship in Roswell, New Mexico.

To special effects experts, the 1947 film, first shown on television in the early 1990s, is an obvious hoax, and not a particularly good one. When the two segments are viewed back-to-back, it's easy to see their point. While the Scientific American segment is in sharp focus, the "1947" segment is often blurry, particularly at key points during the autopsy. And by comparison, the Roswell "alien" looks like a cheap tailor's dummy, with no weight or mass.

"Alien Autopsy" fanned the flames of alien fever that millions of Americans have caught, creating good business for the small town of Roswell. Tour guides lead scores of eager tourists to the Foster ranch, where the "flying saucer" wreckage was first discovered. Back in town, there are gift shops and souvenir vendors on every corner.

Tracing the chronology of the Roswell incident, Alda explains how several historical circumstances, social forces, and examples of faulty reasoning combined over the years to perpetuate the myth of the alien landing. First, archival newsreels of the 1940s and 1950s reveal that flying saucer scares were all the rage, triggered perhaps by the 1947 report of a West Coast pilot who saw some "strange lights" during a flight. In June of the same year, a team of NYU scientists launched several weather balloons in Alamogordo, New Mexico, experimenting with new ways to lift microphones into the stratosphere, hoping to pick up signs of the first Soviet A-bomb tests. The team lost only one balloon, which drifted downwind 70 miles northeast, tracked by ground radar and a B-17 to monitor radio signals. When the balloon's batteries failed, the plane turned back, leaving the research team unaware that the balloon eventually crashed some 15 miles from the Foster ranch, where it was discovered 10 days later by a ranch hand.

In the cold war climate of the West Coast saucer scare, the rancher wondered if he had found the wreckage of an alien spacecraft. It seemed plausible, given the strange-looking electronic equipment, reflective material used to bounce radio signals, and fleshlike neoprene material of the weather balloon. Not taking any chances, the rancher told the local sheriff, who called the nearby army air base. Convinced they had found a saucer, the intelligence officer gave the local press a dramatic story.

The wreckage was flown to 8th Army headquarters at Fort Worth, where meteorologists immediately recognized the materials as parts from a weather balloon. Although General Ramey quickly released another statement to the press to dispel the saucer myth, two decades later a series of paperback books reopened the Roswell incident, this time alleging an elaborate government coverup of the truth.

Charlie Moore, one of the original scientists who worked on the balloon project, gives several examples of how the public's newfound fascination with UFOs contributed to the hoax. For one, the rancher who discovered the balloon debris reported finding bits of plastic tape stamped with a flower pattern. Although decorated tape was merely a popular design of the day and was often used to repair the paper-thin reflective material of the balloon's reflector, confirmation bias may have led the public to misinterpret the rancher's description as evidence of an alien presence. Indeed, a replica role of transparent tape stamped with "alien hieroglyphs" is a popular souvenir that can be purchased in Roswell today.

Aerospace engineer Phillip Class, who has worked for years to debunk UFO incidents such as Roswell, offers another example of how faulty human reasoning has kept the myth alive. Class quotes from several government documents written in 1948 that clearly indicate there was no coverup, noting that these reports have been completely ignored by those eager to keep the myth alive. This is a clear example of *belief perseverance,* the tendency to cling to a belief even after the basis on which it was formed has been discredited.

In closing, the segment notes the importance of Occam's razor in scientific reasoning. This doctrine holds that the simplest explanation is usually the right one. Regarding the Roswell incidence, this of course means that the only aliens around are from Hollywood.

This segment provides a captivating supplement to your discussion of human irrationality, especially as it pertains to social influence and obstacles to problem solving.

Topics for Discussion

- the irrationality of human thinking
- belief perseverance and confirmation bias
- Occam's razor
- social influence

Discussion Questions

1. What explanation of the Roswell UFO incident is offered in the video segment? How did the myth get started? Why did it resurface during the 1970s?
2. Discuss the impact of social influence and irrationality in human thinking in keeping the Roswell myth alive.
3. What is Occam's razor? How would the application of this doctrine dispel the Roswell myth? How do the phenomena of confirmation bias and belief perseverance prevent people from following this doctrine in their everyday problem solving?

Segment 4

Return to the Wild (9:56)

Two hours north of Rio de Janeiro, Brazil, lay the remnants of a forest that once stretched for hundreds of miles. Most of the land is now pasture, most of the animals cows. But now in this narrow strip of forest the trees are home to a thriving population of golden lion tamarins, a species of monkeys once on the brink of extinction. What's unique about these monkeys is that almost all are the descendants of animals born in zoos.

Visiting the National Zoo in Washington, D.C., host Alan Alda is amazed by the unlimited freedom of the colony of tamarins living unfettered in the trees. "At night do they sleep under lock and key?" he asks.

"No," replies director Benjamin Beck. Asked why the monkeys don't wander off, he replies, "Why don't you go to Pittsburgh? They've got their shelter here, they've got their mates here, and they've got their food here. There's no reason for them to go anywhere else."

Beck coordinates a program that takes golden lion tamarins born in zoos around the world and reintroduces them to the Brazilian forest. But before the monkeys are returned to the wild they spend time at the National Zoo learning a variety of survival skills including how to find insects and grubs, even how to peel bananas!

Observing the difficulty the zoo-raised monkeys have in finding their way through a food obstacle course, Beck explains that animals that grow up in zoos typically have weak spatial mapping and locomotion skills. Compared to their counterparts born in the wild, zoo-reared monkeys don't know how to efficiently get around in challenging environments. That's where the National Zoo's training program comes in, functioning as a sort of "jungle boot camp."

After 14 years of running the program, however, Beck was startled to find that the training program did not confer a survival advantage on the tamarins released from the National Zoo. The monkeys were no more likely to survive after special training than were monkeys released straight from other zoos. So, the program now includes a second phase in which the training continues in the wild, allowing the monkeys to learn on their own in the very environment in which they must survive. Food is provided, and one member of each group wears a radio transmitter so that team members can monitor its location at all times. So far they've released 147 monkeys from zoos all over the world. The goal is to build up a population of 2,000 or so monkeys—large enough to eventually become self-sustaining.

Admitting his bias against traditional zoos, where animals are caged and often seem stressed, Alda is pleased that zoos seem to be changing. Beck agrees that over the past 25 years zoos have metamorphosed from menageries that exploited animals to organizations that are seriously committed to conservation, the breeding of endangered animals, and training for wildlife managers.

This segment can be used as an entertaining supplement to your discussion of animal intelligence. It also provides an excellent example of the interaction of genes and environment—nature and nurture—in behavior.

Topics for Discussion

- nature and nurture of behavior
- animal intelligence
- animal learning

Discussion Questions

1. What difficulties do zoo-bred golden lion tamarins face when they are reintroduced to their natural habitats? How does the National Zoo's "jungle boot camp" program attempt to correct these difficulties?
2. In what ways is the mission of today's zoos different from that of zoos 25 years ago?
3. Using examples of golden lion tamarin behavior, explain the meaning of the text's statement, "Nurture operates on what nature endows."

Segment 5

Mind Reading (9:00)

When an otherwise healthy child falls two or more years behind in reading level, the diagnosis is usually dyslexia. The nature of this disability has been fiercely debated for many years. This segment profiles the research of Paula Talal, who believes that dyslexics have a specific disability in processing sensory information.

Sound spectrographs of speech reveal that the sound changes involved in the difference between "puh" and "buh" last only a few hundred milliseconds. Talal believes that these rapid changes are too brief for dyslexic children to recognize.

To prove that the problem for a dyslexic person is not in the ability to hear, Talal devises a test that shows when information is presented to a dyslexic person too quickly, he or she just can't keep up. Talal notes that ". . . the problem seems to be a specific problem that has to do with the rate at which the nervous system can transmit information that keeps coming at it one after the other."

At the National Institutes of Health, other researchers are investigating the physiological bases of dyslexia. As dyslexic subjects are tested for their ability to detect rhyming words, PET scans capture pictures of cerebral blood flow, zeroing in on areas that are responsible for language processing. Some of these areas, such as one in the left parietal lobe, show abnormally low blood flow and neural activity.

Because results such as these imply that dyslexia may have an organic, genetic origin, researchers are developing an early detection test for infants. Early detection will help young dyslexic students receive sustained and supportive teaching to help them become successful readers.

This segment contains material that pertains to several topics, including the links between biology and behavior, the methods used in studying the brain, developmental disorders, and language processing.

Topics for Discussion

- the links between biology and behavior
- PET scan methodology
- reaction time and other methods of investigating infants' perceptual capabilities
- the symptoms, mechanisms, and treatment of dyslexia

Discussion Questions

1. What is dyslexia? What are the criteria for its diagnosis?
2. What has Paula Talal's research revealed about the underlying cause of dyslexia?
3. What is a Pet scan? How does it work, and what does it reveal about the brain? What has this methodology revealed about the brains of dyslexic people?

Segment 6

Image-Guided Surgery (13:07)

The field of medicine is being transformed by the computer's ability to process vast amounts of information.

The central figure in this poignant segment is Linda, who for eight months has had daily seizures caused by an inoperable tumor pushing against her brain. Unless the tumor can be removed, Linda will almost certainly die within two years. With the help of a computer to pinpoint the location of the tumor, Boston neurosurgeon Peter Black believes the operation is possible.

Black begins by developing a three-dimensional model of Linda's brain using magnetic resonance imaging (MRI). MRI is giving physicians unprecedented power to see inside their patients both before and during surgery.

Black discovers that Linda's tumor is not in the motor cortex. "This tells us that it's reasonable to go ahead," notes Black, "assuming that the tumor is now behind the motor strip and not in it. If it was in it we would say that this would not be a possible operation."

Linda is sedated but awake during the actual operation. A video camera superimposes the virtual image of her brain over that of the actual brain. When the images are perfectly aligned, the tumor's location is dramatically revealed.

During the operation, a heavily sedated Linda speaks with segment host Alan Alda. "I went to a lot of doctors. Everyone refused to do this kind of operation; said it couldn't be done. That's why when Dr. Black wanted to do it on television I didn't have a problem. If it could help one other person besides me it would be great."

This segment vividly illustrates how new methods of studying the brain are revolutionizing medicine, and, in cases such as this one, are saving a life by providing a definitive treatment that was once not available. Although the segment presents an uplifting story of the triumph of science over disease, you might alert your students that some of the surgery footage is quite graphic.

Topics for Discussion

- neuroscience
- sensory and motor cortex
- magnetic resonance imaging (MRI)

Discussion Question

1. What is MRI? How does it work? How was it used to help the patient in this video segment?

Segment 7

Severed Corpus Callosum (10:00)

At Dartmouth University, Michael Gazzaniga has been working for many years with Joe, an epileptic who had his corpus callosum severed to stop daily seizures. Cutting the corpus callosum prevented the spread of the seizures from Joe's right hemisphere to his left, but it also prevented the two hemispheres from communicating with each other.

Gazzaniga outlines his classic "split-brain" research paradigm for segment host Alan Alda, providing an especially clear explanation of its rationale. The left half of the brain controls the right side of the body, information on one side of the visual field projects to the opposite hemisphere, and so forth.

As Joe stares at the computer screen, words are flashed to one side of his brain or the other. When a word is flashed to his left brain, Joe can verbally report what he saw. But when only his right hemisphere sees the word, Joe is unable to verbalize what he saw. Using his left hand, however, Joe is able to draw a picture of the object. Not until Joe sees what his left hand (right brain) has drawn is he able to name what it is that was flashed.

"It's almost as though someone has given him a secret communication," notes Alda. "The communication is not occurring inside his head," notes Gazzaniga. "It's occurring out on the piece of paper."

Gazzaniga believes that one of the greatest strengths of the left hemisphere is the drive to interpret why two events occurred. "You can imagine that a species that has that little chip in its brain that asks those questions is going to survive rather well because it's going to figure out more about the nature of the world than a species that doesn't have that."

In a 1990s variation of Gazzaniga's pioneering studies of hemispheric specialization, Joe is shown photographs of paintings by the sixteenth-century artist Archimbaldo, who made faces out of fruits, flowers, even books. Gazzaniga wondered whether the paintings would look different to each of Joe's hemispheres. The first painting goes to the right hemisphere, followed by two words—*face* or *fruits.* He points to *face.* The next painting goes to the left hemisphere, and Joe reports *fruits.* When the image is projected to the left hemisphere the split-brain patient focuses on the elements that made up the face. But when the same image is projected to the right hemisphere, the patient focuses on the face and not the elements.

One side of the human brain is apparently specialized to detect an upright face. "That's right," notes Gazzaniga. "It's an adaptation that we have to detect upright faces. You can imagine in an evolutionary time that all of a sudden you have to quickly detect a face, you want to read an expression on that face, to know if it's friend or foe . . ."

This segment is an effective supplement to your discussion of the structure and functions of the cerebral cortex.

Topics for Discussion

- cerebral structure
- corpus callosum
- hemispheric specialization
- split-brain research

Discussion Questions

1. What is a "split brain?" When is it created, and why?
2. How did Michael Gazzaniga discover the specialized functions of the brain's left and right hemispheres? Explain the rationale of the split-brain experiment.
3. For what specific abilities do the right and left hemispheres of the brain seem to be specialized? Does specialization make sense from an evolutionary standpoint? Explain your reasoning.

Segment 8

Old Brain, New Tricks (11:00)

At the University of Oregon, neuroscientist Helen Neville records electrical activity in the cortex to find out where in the brain we process language. Averaged across hundreds of subjects, her research shows that vocabulary words seem to be in different places in both left *and* right hemispheres. The processing of grammar words, however, seems to be concentrated in the left.

When the study is repeated with young children, who are in the early stages of language development, the results are very different. Young children process language all over the brain. Not until age 4 or 5 does the typical adult-specialized pattern emerge.

This finding may have important implications for how children are educated. "We don't know when the critical time windows are when learning math, learning music, would be optimized," notes Neville, "but I don't have any doubt that there are such critical windows of opportunity. What we do know is that from the point of view of language learning, early is better."

For people who learn a second language, the early locking in of the brain's language areas is especially important. Arthur, who speaks fluent Chinese and English, began learning English as a second language at age 3. When he's tested, his brain's response to English is identical to that of a native speaker—with the same specialized areas for grammar and vocabulary.

Although English is also Nick's second language, he didn't start learning it until age 10. When responding to English, his brain reveals no specialized grammar area.

"The sound and the grammar of a language are the parts that suffer most from delayed learning," notes Neville. "So," she says to Alda—who began learning French as a teenager—"you probably speak with an accent, and your grammar probably isn't perfect. On the other hand, you probably have a huge vocabulary."

In the final story of the segment, Dean Gable, who has been deaf since he was 4 years old, has his brain responses to signs tested. His responses are just like those of any person who learns an early language, with typical grammar and vocabulary areas. But Dean's brain response also indicates that he's using both the normal vision processing area and large parts of an area normally devoted to sound in hearing people. This makes sense, since sound processing is unimportant to a deaf person. This indicates that the developing brain is not only locking in specific functions, but can also invent new uses for sections if necessary.

This segment provides another effective supplement to the latest research questions within the burgeoning field of behavioral neuroscience. It also discusses brain plasticity, developmental issues in second language learning, and the concept of critical periods in development.

Topics for Discussion

- neuroscience
- electroencephalograph
- cerebral specialization in language
- second language fluency
- critical periods in development

Discussion Questions

1. How is neuroscientist Helen Neville investigating brain processes in language? What has she found?
2. Why is it easier for a child to master a second language than it is for an adult?
3. What is a "critical period" in development? What critical periods are described in the video segment? Can you think of other critical periods in human development?

Segment 9

Smart Glasses (8:13)

This segment explores how perception researchers are helping blind persons by developing the equivalent of "bionic senses."

Leonard is a victim of macular degeneration. To get by he carries a variety of magnifying lenses: a small telescope that magnifies things until they are legible to his peripheral vision and a low vision lens for near objects.

But now Leonard is being fitted with a prototype system, called "Elvis," that used advanced video and computer technology to provide a much clearer (and less cumbersome) view of the world. The system uses a video camera that processes the visual field electronically.

Elvis was designed by Bob Masov of the Wilmer Eye Institute. Masov is working on a refinement of the system that will help even those with the poorest visual acuity.

Masov demonstrates how many visually impaired people, who are unable to see sharp detail, can decode images that provide high contrast—such as a caricature of a person's face. His new system consists of a computer program that converts an image into a high-contrast caricature. In addition to the perceptual advantages it offers, the software can operate in real time, as a visually impaired person is viewing the world.

This segment makes a nice supplement to your coverage of vision and visual perception. It also vividly illustrates how the fields of applied psychology and artificial intelligence have tackled a major human health issue.

Topics for Discussion

- visual system
- visual disorders
- bionic senses
- visual perception
- feature detection

Discussion Questions

1. What is macular degeneration? How is it "corrected" in the video segment?
2. How has Bob Masov used computer processing to improve the perceptual ability of visually impaired people? What does the success of this system demonstrate about visual perception?

Segment 10

Lights, Camera, Magic! (12:53)

Director Michael Owens and his team from Industrial Light and Magic—famous for their work on the *Star Wars* trilogy—are magicians when it comes to creating special effects. An army of model makers, painters, puppeteers, and countless other specialists create Godzilla.

The lighting is set, and the 6 foot, 6 inch Charles Barkley, who is supposed to appear 160 feet tall in the commercial, is ready for his first scene. To make him look taller, a miniature city has been built. But the set is only part of the illusion. Camera angles, special film speeds, hand-drawn animation, radio-controlled puppets, blue screen, and video layering are just a few of the other tricks Owens will use to make the video convincing.

Eventually, the video comes together as a single montage of flames, falling billboards, fire works, helicopters, and fleeing people.

After eight days of filming and four weeks of editing, the 30 seconds of wizardry is ready for prime time. The segment ends with the completed commercial, as it appeared on the air.

This segment can be used as an entertaining supplement to your discussion of visual perception. It provides an excellent example of the involvement of many basic perceptual processes in film.

Topics for Discussion

- perception of size, movement, depth
- perceptual illusions

Discussion Questions

1. What are some of the many visual illusions that appear in the Charles Barkley Nike shoe commercial?
2. In the making of the Nike shoe commercial, technicians from Industrial Light and Magic capitalized on several basic perceptual processes. What are they?

Segment 11

Cockpit Confusion (11:14)

Modern jet liners are highly automated. Computers operate the controls, run the engines, navigate, and make the landing. But automation causes its own problems. These problems, and the solutions offered by human factor psychology are the subjects of this segment.

MIT engineers study cockpit display design to help eliminate "mode confusion" in the cockpit. The computer can steer the plane in "navigation mode," by flying a specific course through precise check points, or in "heading mode," simply by having the plane follow a preset heading. The display that differentiates these modes is very clear; it was well-engineered to prevent possible human error.

In contrast, the display for vertical navigation is much more complex. The controls allow the computer to select a particular altitude, maintain a given height, or climb or descend at a set rate. But there's no simple visual display of what the plane is doing; it's all knobs and numbers.

Vertical navigation confusion apparently led to the 1988 Airbus crash in Strasbourg, France. Forced at the last minute to land on a secondary runway, the pilot had to quickly reprogram the computer with the new descent angle. But depending on its position, the programming knob controls descent speed as well as angle. Also, the display is confusing. The only difference between a 3.2 degree descent angle and a descent speed of 3,200 feet per minute is the presence of a dot between "3" and "2." The pilots thought they had set 3.2 degrees, when in fact they had called for a 3200 ft/minute descent speed. This is a much steeper angle, which caused the plane to crash into the mountain.

Although this clearly was a case of pilot error, it is also a case of poor human factors engineering. After the crash, the airline improved the display, but the task of improving communication between pilots and computers isn't complete.

In a second example we see how pilots must constantly be on guard to make sure computer controls don't actually get the plane into trouble. In certain landing situations, for example, the computer may actually change the altitude at which the pilot has instructed the plane to approach the runway. If there's other traffic at that altitude, there could be a crash. Most pilots will anticipate this and by-pass the computer. But in complex, real-world situations, when, in addition to paying attention to landing instructions and the computerized controls, pilots are getting weather information, dealing with passenger problems, monitoring fuel, and so forth, they may overlook this crucial adjustment.

Human factors scientists are working on new displays to correct this type of problem. Compared to the old days, computers have made flying a lot safer, but they have also made being a pilot a different kind of job. Today, safe pilots are pilots who know how to manage the automation.

This segment is an excellent supplement to your discussion of applied psychology, especially human factors psychology.

Topics for Discussion

- human factors psychology
- sensory overload

Discussion Questions

1. What is human factors psychology? What kinds of issues are addressed by this new subfield of psychology?
2. What apparently caused the 1988 Airbus crash in Strasbourg, France? What changes in cockpit design were initiated following the crash?
3. What are some other applied problems that human factors psychologists should investigate?

Segment 12

Tasters and Supertasters (14:00)

In Santa Fe, New Mexico, host Alan Alda visits a food festival to celebrate what's practically the state food—the Chile pepper. An avowed hot pepper enthusiast, Alda wants to know what makes them taste hot. Paul Bosland of New Mexico State University explains that the heat comes from a set of compounds called *capsaicins.* In nature, peppers evolved these spicy compounds to keep mammals from eating the fruits.

In Baltimore, Alda pits his Chile pepper sensitivities against those of a panel of experts at the McCormick Spice Company. Starting with a mild solution of capsaicin, the panel is instructed to assign it a "5" on a heat scale ranging from 0 to 15. With their tongues calibrated, the panel samples a real hot pepper. The panel, including Alda, displays remarkable consistency in rating this much hotter stimulus, with testers assigning values of 8, 7.9, 7.5, 7.2, 7.6, 8.2, and 7.8.

At Yale Medical School, Linda Bartoshuk gives Alda a taste test, confirming that he, like one person out of every four, is a "supertaster." Bartoshuk places a chemical-saturated paper disk on Alda's tongue, which is then painted with blue food coloring. The food coloring reveals that Alda's tongue has an abundance of fungi-form papillae, each of which harbors a half dozen or so taste buds with nerve fibers connecting to his brain.

Whereas most fungi-form papilla are, in fact, taste receptors, many don't sense taste at all, but rather respond only to pain. Capitalizing on this property of Chile peppers, Linda Bartoshuk makes candy laced with cayenne pepper to treat people suffering from painful mouth sores. Bartoshuk suspects that every culture that has enjoyed Chile peppers has eventually discovered their analgesic properties.

Today, capsaicin is also used to relieve more serious forms of chronic pain, such as the neuropathy suffered by long-term AIDS survivors such as Gepetto, an avid runner who began suffering excruciating pain in his feet several years after testing HIV-positive. Unable to run, Gepetto became housebound, with his pain made only controllable with the help of powerful tranquilizers. Anesthesiologist Dr. Wendye Robbins reasoned that if hot peppers numb pain in the mouth, why not elsewhere? After all, the same nerve fibers that signal hot or spice in the mouth are also present in the foot and probably could be activated in the same way.

At the University of California, San Francisco, David Julius has isolated the molecule in our nerves that is activated by the heat of hot peppers. The molecule sits like a trap door on the surface of the pain fiber. Capsaicin unlatches the door, allowing calcium ions to rush in, firing off the pain messages to the brain. Capsaicin fools cells into thinking they are on fire. Paradoxically, once the burn wears off, the result is relief.

The result of these programs of clinical research was the recent invention of capsaicin-based pain-relieving creams. The treatment had Gepetto back running within a week—a benefit that may last as long as several months, and one that doesn't rely on dangerous painkilling drugs.

This segment is an appropriate supplement to your discussion of sensory processes, receptor physiology, and pain. It is also a striking example of how good scientists reason their way into important discoveries; in this case, a discovery with far-reaching clinical applications.

Topics for Discussion

- sensation and perception
- taste receptors
- pain
- evolutionary perspective

Discussion Questions

1. What makes Chile peppers taste hot?
2. Working from an evolutionary perspective, explain why Chile peppers produce a sensation of heat and pain in mammals but not in birds.
3. In what sensory and perceptual ways do "supertasters" differ from "tasters" and "nontasters"?
4. Explain the reasoning behind the invention of capsaicin pain-relieving creams, and describe the physiological effect of such creams on pain nerve fibers.

Segment 13

Catching Catnaps (11:45)

Solo sailors quickly discover that they must drastically cut back on the amount of sleep they're used to getting in order to perform countless tasks during an extended ocean voyage. But how much can the body adapt? Is it best to take it in one period of slumber or to spread it out among multiple catnaps?

Attempting to answer these questions, researchers at a Boston sleep institute persuaded an Italian artist, Francisco, to cut his daily amount of sleep from 8 to 3 hours distributed over 6, half-hour naps. The naps were sandwiched evenly between waking periods throughout each day over a 7-week period.

Researchers compared Francisco's "normal sleep" polygraph reading to his fast-wave, slow-wave, and REM responses once his sleep regimen was altered. What type of sleep will the body choose when it is allowed to sleep for no more than 30 minutes at a time?

As the experiment proper begins, Francisco shows few adverse effects. His self-reported alertness and fatigue, dream recall, and performance on various cognitive and motor tasks indicate that he is having little difficulty adjusting to the experiment.

Physiologically, however, the picture is quite different. Although the polygraph indicates that the three types of sleep are still present, the architecture of Francisco's sleep has changed. Normally, the first REM episode begins 90 minutes to 2 hours after the onset of sleep. After only a few days of the experiment, however, Francisco enters REM almost immediately after falling asleep. As the study progresses, it soon becomes apparent that although total sleep has been reduced dramatically, the percentages of sleep time allocated to fast-wave, slow-wave, and REM are remarkably stable. Francisco's body has made an extraordinary adaptation in its sleep pattern in order to preserve the relative amounts of each type of sleep.

By day 48, however, even Francisco's girlfriend has extreme difficulty awakening him at the end of each 30-minute catnap. It seems that after two months of sleeping only 3 hours a day, there is a tremendous build up of "sleep pressure."

As Francisco cuts his tether to the polygraph on the final day, it's clear that Francisco is glad his ordeal is over. "I'm ready for a little vacation," he says.

This segment is an appropriate supplement to your coverage of several content areas, including research methodology, states of consciousness, and health psychology.

Topics for Discussion

- laboratory versus field research
- limitations of case study and self-report studies
- single-blind control procedures
- states of consciousness
- electrophysiological measurement (EEG, EMG, EOG)
- sleep stages
- human adaptability to altered sleep schedules (shift work, jet lag)

Discussion Questions

1. What are the three basic types of sleep that are differentiated in this module? What physiological changes occur during each type of sleep?
2. How did Francisco's sleep change in response to the multiple-naps regimen? Were there any adverse cognitive or psychological effects?
3. What research method is this study based on? What are several advantages (and limitations) of this method?
4. What conclusions regarding sleep did the researchers draw from the results of their study? Why should you be cautious in generalizing from the results of this study?

Segment 14

What's in a Dream? (13:00)

Alan Alda joins a Harvard University's sleep research project as a sleep subject. The study involves finding out what happens to the mind while we are dreaming. The experiment begins with a simple cognitive reaction time task. By measuring how long it takes to compare two words, the researchers have an index of how good the brain is at making associations.

As Alda falls asleep, the polygraph records his brain waves and eye movements. Several times during the night he is awakened and asked to perform the word association task. He's also asked to report his dreams when the polygraph reveals that a REM episode has occurred.

Most dream researchers believe that during REM sleep the normal signals from the brain are cut off. Instead of receiving inputs from the eyes and ears, the visual and auditory centers are flooded with signals surging upward from more primitive regions of the brain. These signals are believed to be random and completely meaningless.

Harvard psychologist Bob Stickgold explains that during dreams our brains are scrambling to make sense of nonsense. And this is where the word association task is revealing. Subjects are faster at making associations following a REM episode than they are when they are awakened from NREM sleep, or even when they are wide awake. It's as if during REM sleep the brain is primed to put together stories from random images and feelings.

Other research conducted at Toronto's Trent University suggests that dreaming helps us learn. College student Lara is spending four nights of her summer vacation in the lab. Researchers record and count her eye movements during a time when she's in a relaxed, relatively stress-free mode. The experiment is repeated four months later during final exam week, when Lara's mind is in a strenuous learning mode.

Note that the pattern of eye movements is strikingly different during dreams. "For some people," notes researcher Carlisle Smith, "there's almost a doubling of eye movements after they've had intense learning activity."

To investigate the purpose of the extra eye movements, Smith sets up an experiment to see if learning a complicated logic game is affected by how much dreaming a person does. After the students are tested, they receive a simple paired-associates memory test. Finally, all of the subjects are allowed to sleep. Some are awakened during REM episodes. Other control subjects are either awakened during NREM episodes or allowed to sleep uninterrupted.

One week later the subjects are retested on the paired-associates and logic tests. Although there were no differences in performance on simple memory task, dream-deprived subjects did significantly worse than control subjects on the more complex logic test.

In a final version of the experiment, students hear a loud clock ticking while they learn the logic game. Later, as they sleep, some wear headphones and hear the ticking during REM periods. The idea is to see if the ticks remind the dreamer of the learning task. Students who heard ticking during REM were far better than control subjects in learning the logic task. To Smith this suggests that being reminded of a problem during dreaming helps us tackle it.

This segment gives an effective overview of sleep stages and outlines the latest theories of the nature and function of dreaming.

Topics for Discussion

- sleep research methodology
- REM and NREM stages of sleep
- mechanisms and theories of dreaming

Discussion Questions

1. Outline the course of a night's sleep, differentiating the body's state during REM and non-REM stages.
2. Why, according to most psychologists, do we dream?
3. How has Bob Stickgold investigated the purpose of dreaming? What has he found?

Segment 15

Can You Beat Jet Lag? (6:44)

Like many travelers who struggle with the disorientation of international air travel, host Alan Alda wonders whether there might be a simple way to beat jet lag. At Cornell Medical College, researchers Patricia Murphy and Scott Campbell believe there might be—one that involves light and the resetting of the body's biological clock.

Since 1980, phototherapists have used large, glaring banks of light to treat people with sleep disorders and other conditions presumably influenced by the body's natural biological rhythms. The conventional wisdom was that a neural pathway linked the eyes to the brain's biological timepiece. A strong visual could, in theory, produce all sorts of beneficial effects by resetting this mechanism to an earlier or a later time.

But Campbell and Murphy believe that the light need not be glaringly bright, or even shone in the eyes at all. Using Alda as a guinea pig, they test their hypothesis that the body may have a second time system, carried by some mechanism in the bloodstream, which serves as a backup to the more direct visual route to the brain.

The researchers strap two pads of a weak blue light used to treat neonatal jaundice to the backs of Alda's knees. These sites were chosen because the back of the knee is a place where the blood runs very close to the surface of the skin. With the light pads in place, Alda watches a three-hour movie and is then asked to provide a sample of saliva, which is analyzed for melatonin, a hormonal marker of activity of the body's biological clock. Along with the results from other subjects in their pilot study, Alda's melatonin level indicates that light to the back of the knee has the same effect as light presented to the eyes. More specifically, by varying when the light stimulus is presented, Campbell and Murphy have been able to reset the clock to an earlier or a later time, with equal effectiveness.

Although the experiment remains to be replicated by other researchers, if confirmed the results may offer hope to people who suffer from jet lag, as well as those who frequently awaken during the night. Like Alda, many people with these problems have biological clocks that give their wake-up signal in the middle of the night. In theory, light therapy resets the clock, pushing the wake-up signal back an hour or two, as needed. But since the best time for the light is just before the wake-up signal is given, conventional phototherapy often doesn't work since most people find it difficult to fall asleep with a bright light in their eyes. The light pads would allow light to be administered at the appropriate time without disrupting sleep.

This segment makes an interesting supplement to your discussion of biological rhythms, sleep-waking cycles, and sleep disorders. It is also a powerful demonstration of how the brain and body adapt to changing environmental circumstances.

Topics for Discussion

- biological rhythms
- sleep-waking cycles
- jet lag
- sleep disorders

Discussion Questions

1. What is a biological rhythm? How are such rhythms implicated in sleep disturbances and jet lag?
2. How is conventional phototherapy used to reset the "biological clock" and treat sleep disturbances?
3. What new, experimental form of phototherapy is introduced in the video segment? What advantages does this form of treatment offer over conventional phototherapy?

Segment 16

Remembering What Matters (8:30)

University of California-Irvine researcher Jim McGaugh has been studying how the mind is shaped by memory. In one experiment, a rat learns which arm of a radial maze contains food. Eighteen hours later, however, the rat has forgotten the task. When another rat is injected with adrenaline just after training, however, the memory remains after 18 hours.

Adrenaline is the hormone responsible for the "fight-or-flight" response. McGaugh's research suggests that the adrenaline rush does more than prepare animals to meet emergencies, however. "It also would be a good idea to remember where the predator was and what happened so that the next time the animal can avoid that situation . . . the same hormones that were involved in generating the fight-or-flight response, we now have discovered, work on the brain so as to make stronger memories."

In another experiment a rat swims to find a transparent underwater platform. Once he's shown where the platform is, the memory is stamped in. Three days later the rat is able to find the platform quickly. But when a rat is injected with a beta-blocking drug just after training, it can't remember where the platform is located. The drug works by blocking the effects of adrenaline. For rats, adrenaline appears to be central to making strong memories.

In the next scene, a human version of the experiment is depicted. A student listens to a boring story about a mother taking her son to visit his father's workplace. When the subject is asked to rate his emotional reaction on a 10-point scale, he responds with "2."

A second subject hears a very different version story. This time the boy is struck by a car as he crosses the street with his mother. When the subject is asked to rate her emotional reaction, she responds with "7."

Two weeks later the subjects' memories of story details are measured. Memories of the emotional story are much better recalled than memories of the boring story.

To determine whether adrenaline is responsible for the differences in memory, the researchers give another group of subjects a beta-blocking pill just before listening to the story. Although the story is still rated as highly emotional, two weeks later memories for details are just as poor as in those who heard the boring story.

PET scans reveal that the amygdala is the most active region of the brain as an emotional memory is formed. The more active the amygdala, the better the subjects' memory for story details weeks later. Activated by the hormones the emotions produced, the amygdala sends a message to the rest of the brain as if to say, "This is important! Don't forget it."

This is an effective supplement to your discussion of memory and emotion. It also provides a nice description of PET scan technology and brain mechanisms in information processing.

Topics for Discussion

- memory, emotion, motivated forgetting
- fight-or-flight response
- neural mechanism of agonists and antagonists
- PET scans
- limbic system

Discussion Questions

1. How has McGaugh's team of researchers investigated the relationship between emotional arousal and memory in animals? in humans? What have they found?
2. What is the "fight-or-flight" response? What aspects of the nervous system and brain are involved in the processing of emotional memories?
3. What evolutionary advantage might have been conveyed to our species by the close connection between emotional arousal and memory?

Segment 17

True or False? (9:00)

Noted memory researcher Dan Schacter gives Alan Alda a simple memory test. Sitting on a park bench, the two watch an actor and actress enjoy a carefully choreographed picnic.

When the play is over, and Alda has left, the scene is replayed and photographs are taken. Some photos depict actual scenes from the play. Others depict events that did not actually occur.

Schacter's premise is that memory is malleable. "One of the things we know about memory is that it's not fixed . . . the way we talk about the event later, the way we think about it. . . can sometimes change our memories later."

Two days later a suspicious Alda is shown the photographs and asked to differentiate between events that occurred and those that did not. "Dan was obviously trying to confuse my memory of things I'd seen for real with things I'd only seen in the photographs."

Even though Alda knew some of the things in the photos had never happened, his memory was confused. Of eight false events, Alda incorrectly remembered two as having taken place. Schacter is also studying how the brain represents real and false memories.

"Instead of being in one place," he explains, "many of us believe that memory is scattered in different parts of the brain. Memory consists of all the bits and pieces of an experience, the sights, sounds, and emotions, with each fragment stored in areas of the brain responsible for handling that particular sensation . . . sounds are stored in the auditory cortex, and so on."

The final segment reviews Schacter's recent PET scan studies differentiating real and false memories. Subjects heard lists of related words that they later were asked to recall. The words in each list are united by a theme word that is not on the list. For example, one list contains the words *bed, doze, nap,* and *yawn. Sleep,* however, is not one of the words on the list.

PET scans revealed that while recalling both true and false memories activated the hippocampus, only true memories activated the auditory cortex. Although the subjects reported hearing the words that weren't there, their brains contained no trace of the sounds of the words.

Schacter emphasizes that there's a long way to go before the trace of a false memory can be used in a practical test that can be used, for instance, in a courtroom.

This segment highlights several issues in the nature of memory. It also describes how neuroscientists investigate brain mechanisms in complex psychological phenomena.

Topics for Discussion

- constructive memory
- brain areas involved in memory
- false memory

Discussion Questions

1. What has researcher Dan Schacter discovered regarding the accuracy of long-term memory?
2. How does the brain store a memory? How is it recalled?
3. How does the brain differentiate between actual and false memories? How did Schacter discover this?

Segment 18

Talkin' Babies (12:00)

Researchers at Montreal's McGill University are studying the earliest hallmark of language development: babbling. Children the world over spontaneously repeat a single syllable over and over, producing utterances such as "mamamama." All babies do this, but according to psychologist Laura Petitto, babbling is not language as we know it—as meaning and content—it's language as a baby knows it—a form of play with forms and sound.

If babbling doesn't mean anything, then why do babies do it? Most linguists think babies babble because they are establishing control over the muscles that produce speech. Petitto, however, thinks babbling has more to do with developing language ability.

This disagreement is fueling an intense debate about the nature of language itself. The central question is whether language and speech are as intimately connected as they seem. The human brain has evolved to work closely with the vocal tract to produce speech. But if the vocal tract hadn't evolved the way it had, would humans still have developed language?

Petitto believes so. According to her viewpoint, language is an independent part of the brain that would find some way to come out even in the absence of sound.

Eighteen-month-old Remy, who along with his parents is profoundly deaf, provides a unique opportunity to test Petitto's theory. If Petitto's theory is correct, that babbling is tied to language and not to speech, then deaf children should babble with their hands.

Laborious videotape analyses of Remy signing indicate that deaf kids do babble. In a second example, 9-month-old Vance, also deaf, is at the dinner table with his sister and his mother. Vance is not yet able to join in the sign language conversation, but tries to get involved by placing his hands directly into their line of sight and making signs that have all the features of vocal babbling. The repeated syllable consists not of a consonant and a vowel, but of a hand shape and a hand movement.

"The babbling is the child's active attempt to master the form of language, to listen to the environment, to look at the environment, to look for a particular structure, to extract out that structure and in little baby steps play with the forms of language in an attempt to build and master a target language."

But is sign language just a substitute, something the brain turns to when speech is not available? To find out, the McGill researchers have been following the development of children such as 2-year-old Simone, who has one deaf parent and one hearing parent. Simone, who has normal hearing, is learning both to sign and to speak. Simone passed every milestone of language learning in both sign and speech at exactly the same time.

This segment provides an informative supplement to your discussion of language, including its structure, acquisition by children, and theoretical origins.

Topics for Discussion

- structure of language
- stages of language development
- linguistic universals in language development
- nature versus nurture in language acquisition

Discussion Questions

1. What is babbling? What does it reveal about language development?
2. Why is the issue of whether deaf babies babble with their hands significant to our understanding of the origins and nature of language?
3. What evidence is offered in the segment that the brain doesn't "care" whether it receives spoken or signed language input as it develops?

Segment 19

If Only They Could Talk (10:07)

The 1995 movie *Babe* expressed an age-old human fantasy—talking animals. This segment illustrates several examples of researchers who are attempting to use human language to communicate with a variety of animal species. In the first example Hamlet, a pig owned by Ohio State University psychologist Sally Boysen, correctly responds to a simple gestural language in which balls, dumbbells, and other objects are each associated with simple signs. After two laborious years of training, Hamlet is able to trot over to the correct ball when Sally gives its sign. Although simplistic, Hamlet's associations are real, as Sally took care to avoid the *Clever Hans effect,* by which her eye movements, gestures, or nods could tip Hamlet off.

Although Hamlet's memory is perfect (even though he hasn't been tested in three years), simple associations are his limit. He can't use the signs in any other way to show that he has any understanding beyond a mere association.

In the second example, Ron Schusterman of the University of California at Santa Cruz is studying Rocky, a sea lion, who has been trained to respond to another gestural language—this one consisting of 24 signs. As with Hamlet, the signs of Rocky's language refer to a variety of different objects. Unlike Hamlet, however, Rocky can also use the signs in several ways, such as responding to simple commands and questions.

Trying his hand at communicating with Rocky, host Alan Alda issues a simple command: "Ball . . . flipper touch . . .," and Rocky instantly swims to the ball and taps it with her flipper. Now Alda asks a question: "Football . . . is it here?" When Rocky heads right for the "yes" button, Alda is pleased. "It's an exhilarating feeling, to really get through to an animal. It's obvious that she really does get it."

In archival footage from the late 1960s we see one of the earliest attempts to teach human language to an animal. Raised from infancy by Alan and Beatrice Gardner, Washoe the chimpanzee learned over 130 American Sign Language signs by the time she was 5 years old.

The final example of the segment is perhaps the most impressive. Alex the parrot can distinguish and name different objects and colors. He can even count. Asked how many keys a researcher holds, Alex immediately responds "two." Most impressive of all, Alex even shows signs of understanding abstract concepts such as sameness and difference. When researcher Irene Pepperberg of the University of Arizona holds up a gray metal key and a yellow plastic key, and asks, "What's different?" Alex hesitates, then states, "color."

And Alex's performance is not a mere circus trick limited only to a small number of objects and concepts that are repeated over and over. He can juggle concepts in an astonishing way. In the next demonstration, Alex is shown a tray of seven different objects and asked "What matter four corner blue?" (Alexspeak for "What is the 4-cornered blue object made of?") There are several objects with four corners, and several blue objects, but only one (a blue square made of wood) with both features. Although he's never been asked this question before, Alex carefully studies the objects and soon comes up with the correct answer "wood."

Despite the apparent successes of these projects, animals are clearly limited in their abilities to learn human language. It's obvious that neither their brains nor their bodies are organized to handle even the simplest human forms of communication. The challenge for researchers is to probe animals' minds without the aid of language, to see if somewhere in them there are, in fact, concepts and ideas.

This segment is an entertaining supplement to your discussion of several content areas, including language and thinking, animal research, and the scientific method.

Topics for Discussion

- can animals exhibit language?
- the Clever Hans effect
- the Washoe study

Discussion Questions

1. In what ways have psychologists attempted to teach animals to use human language? How successful have these projects proven to be?
2. Why are the accomplishments of Washoe the chimp, Rocky the sea lion, and Alex the parrot linguistically more impressive than that of Hamlet the pig?
3. What is the Clever Hans effect? How might this effect lead an unsuspecting researcher to draw erroneous conclusions regarding an animal's ability to exhibit language? How do researchers guard against this form of confounding in their animal research?

Segment 20

Who Needs Words, Anyway? (10:53)

This captivating segment illustrates another, nonlinguistic approach to the study of animal intelligence: experimental procedures designed to determine whether animals can learn to reason abstractly.

At the University of California at Santa Cruz, Ron Schusterman is working with Rio, a sea lion that he claims can understand the abstract concepts of grouping, or classification. Shown two nonmatching symbols, Rio is reinforced for consistently choosing letters or, on other trials, numbers. Learning the task was an arduous process that took Rio thousands of conditioning trials and more than a year of training.

After a flawless performance on a series of choose-the-letter trials, the reinforcement suddenly stops. Clearly frustrated, Rio takes a quick swim in her tank. On the very next trial she starts picking numbers, and she continues doing so as long as she's reinforced. As a twist, Ron adds a number that Rio has never seen before. But Rio has no trouble, apparently using the principle of exclusion to identify the "nonletter."

On a third series of trials Rio is shown a letter or number in the center of a stimulus array and reinforced for choosing a matching letter or number from a pair of symbols presented on either side. Rio's performance on the "match-to-sample" procedure is once again flawless, leaving little doubt that she has demonstrated a rudimentary understanding of the principle of grouping.

Although Schusterman can only speculate how sea lions in the wild might use the ability to sort objects into abstract groups, he thinks that classification might help them sort out the many individuals they encounter, labeling them as friends or foe, family and nonfamily, and so on.

Ohio State University psychologist Sally Boysen is convinced that chimpanzees also can learn abstract concepts with ease. One of her subjects is 17-year-old Sheba, whom she raised from the age of 2. Sheba watches closely as Sally hides a dollhouse-sized can of Pepsi in a tiny storage cabinet of a scale model of an adjoining room. The scale model is filled with a variety of miniature objects and pieces of furniture, each of which has a full-scale counterpart in an adjoining room. While Sheba stays in the other room, Sally hides an actual can of Pepsi in the full-size storage cabinet in the adjoining room. When she is later given the opportunity to enter the full-scale room, Sheba immediately locates the can of Pepsi in the storage cabinet.

As simple as this symbolic representation ability seems, University of Chicago psychologist Judy DeLoache points out that it eludes human children until about 3 years of age. Using the same procedure as Boysen, DeLoache poses the hide-and-seek task to 2 ½-year-old Amos, who just doesn't seem to get that the model symbolizes the real room. So in one way, Sheba is sharper than the average 2 ½-year-old child. There's no doubt that Sheba sees the model as a representation of the real room. "If that's not abstract thinking," notes host Alan Alda, "I don't know what is."

This segment is an appropriate supplement to your coverage of several content areas, including animal intelligence, cognitive development in children, and evolutionary psychology.

Topics for Discussion

- animal intelligence
- cognitive development
- abstract reasoning

Discussion Questions

1. In what ways have psychologists investigated abstract reasoning in animals? How successful have these projects proven to be?
2. Explain the logic behind the match-to-sample and scale-model test procedures used by Ron Schusterman and Sally Boysen. What does successful performance on these tasks demonstrate about the cognitive abilities of young children, sea lions, and chimpanzees?
3. How might a psychologist working from the evolutionary perspective account for a sea lion's ability to understand an abstract concept such as grouping, or classification? What adaptive significance might this ability have for animals in the wild?

Segment 21

Born to Talk (6:45)

This segment begins with a reenactment of Jean Berko's classic study of grammar acquisition in children. Psycholinguist Steven Pinker teaches 3-year-old Peter a new word.

"This is a wug. I've got another one, now I have 2 ___?"

Peter has no trouble responding ". . . wugs!"

Pinker and his colleagues at MIT have taught nonsense words to hundreds of kids to see if they will apply the same rules of grammar to made-up words as they do to real ones. The results have led them to conclude that children are born with an instinct for acquiring language.

"Human language is very special," says Pinker. "Kids don't have simple sentences prefabricated in their brains. They put them together on the fly." The reason they are able to do so, according to Pinker, is that they are born with an innate understanding of grammatical rules.

To convince host Alan Alda that this is true, Alda is asked to teach a child how to "chan" with a toy Cookie Monster. (Alda flips the toy from a teeter totter.) Afterward he asks the child "what did I just do?" Without hesitation she says, "You channed with Cookie Monster."

In the next scene we see another universal feature of language acquisition: the mistakes children make when they overapply grammatical rules. After Alda has acted out a story with stuffed animals, Erin is asked to retell the story. Doing so, however, requires her to form the past tense of several irregular verbs. How will Erin handle this problem? She does so by over-regularizing verbs such as "draw" and "stick." As she comes to the appropriate points in the story, without hesitation she adds "-ed" and says "drawed" and "sticked."

"That's another way we know that they are not just memorizing words," notes Pinker. "Right from the beginning they are abstracting out rules and applying them to new forms. . . . 'cause if they say 'bringed' they haven't heard mom and dad say that, mom and dad say 'brought.'"

"Obviously no language is innate. Take any child of any race and bring him up in any culture and they will learn the language equally quickly. But what might be in the genes is the ability to acquire language—not any particular language, but the bits and pieces that they are born with."

This segment provides a clear replication of several classic studies of language development in children. It also touches on the nature-nurture controversy as it applies to cognitive development, vividly differentiating the position of behaviorists like B. F. Skinner from that of Noam Chomsky, who argued for the existence of an innate language acquisition device.

Topics for Discussion

- stages of language development
- linguistic universals
- over-regularization
- theories of language acquisition

Discussion Questions

1. What is Steven Pinker's theory regarding language acquisition? How has he tested this theory? What has he found?
2. Why are the grammatical mistakes that all children make as they acquire language important to psycholinguists? What do they reveal about the origins of language?

Segment 22

Bypass Genes (8:09)

Lillian Cooper used to walk 5 miles every morning at a Natick, Massachusetts, shopping mall. But she has recently been sidelined by a badly narrowed artery in her left leg: "If I don't find a way to get it fixed, I'm going to lose the leg. I'm not ready for that."

Lillian has already tried all the standard therapies for her blocked artery. Restoring blood flow in a blocked artery is usually attempted with bypass surgery or balloon angioplasty. But for Lillian, these techniques have already failed.

Lillian's doctor is recommending gene therapy—an experimental technique that many believe will revolutionize medicine in the twenty-first century. Although the field is still in its infancy, researchers are hopeful that the Human Genome Project and its efforts to map the 100,000 or so human genes, will pave the way for future successes. In Lillian's case, the idea is to see whether putting a particular gene into blocked arteries will cause them to develop shoots that will bypass the blockage. Notes Lillian's doctor, "There's nothing like letting nature do the surgery."

If the gene therapy grows a bypass for Lillian's blocked leg artery, there are obvious medical implications, including the possibility of similar procedures on coronary arteries. In fact, the ultimate goal is to use gene therapy as an alternative to bypass surgery for people with heart disease.

Four weeks after the treatment Lillian is more hopeful than ever. Walking the one-half mile to the clinic without aid, she beams, "My leg is better, my foot is better. I feel that there have to be new blood vessels forming." Another angiogram partially justifies her optimism. Before surgery, it took 15 seconds for blood to reach her calf; now it takes only 9 seconds.

Gene therapy is still in its infancy. Most of the 100 or so clinical trials have been disappointing or inconclusive. "A lot of the things we try turn out to be science fiction . . . make good movies but they don't help too many patients. . . . Now we're seeing a few indicators that this might actually be therapeutic for certain groups of patients."

This segment can be used to supplement your discussion of the mechanisms of heredity. It can also be used in classroom discussions regarding the advancement of medical technology and the increasingly controversial issue of medical ethics.

Topics for Discussion

- mechanisms of heredity
- the Human Genome Project
- gene therapy

Discussion Questions

1. What is gene therapy? How was it used to treat the patient depicted in this video segment?
2. As new technologies develop, many controversial/ethical questions emerge. For example, should prospective parents be tested to determine if they carry any potentially serious genetic abnormalities? How do you feel about this issue? Can you think of other issues that are raised by advances in medical technology?

Segment 23

Bringing Up Monkey (9:40)

Each person has a unique way of handling the challenges of life. Some of us are bold and confident; some a bit more timid and uncertain.

For the past 10 years, researcher Steve Sumi has been studying a colony of 30 rhesus monkeys, who, like humans, display striking individual differences in personality. What Sumi and his team of researchers have discovered is that these different types of personalities are strongly associated with the type of relationship that exists between mothers and their offspring. The boldest monkeys have a *secure attachment* to their mothers—they play confidently, periodically returning to mom when they are frightened. Their mothers are relaxed and don't hover when their offspring venture away to play.

The relationship between timid monkeys and their mothers is very different. Timid monkeys cling to their nervous mothers (*reactive* style), who seem overly protective of their babies.

Remarkably, the personality differences in monkeys are apparent in the first days of life, long before environmental experiences and learning could account for them. The researchers are convinced that the personalities of rhesus monkeys probably aren't learned from parents; they are inherited through genes.

But the researchers wonder how rigidly temperament is fixed. To find out, research is conducted by placing a reactive newborn rhesus monkey with a relaxed foster mother.

The results? The foster mother taught the reactive baby to be more relaxed. He's now showing all the signs of being a bold, young monkey, happy to play and explore on his own.

But 4 months later, the monkey is separated from his foster mother, and he becomes upset and obviously distressed. Despite being raised by a foster monkey who is calm and nurturing, the monkey has reverted back to being a very highly reactive individual. It seems that in extreme situations, the genes win out.

"High reactive infants that grow up in benign environments won't have any trouble at all," notes Sumi. "High reactive infants that grow up in environments that are full of stress and challenge will need help if they want to make it through in a successful fashion."

The narrator speculates about the relevance of Sumi's findings to our understanding of human development. It seems safe to conclude that, despite inherited personality traits, a parent's behavior toward his or her child has a real effect, within limits, on that child's attitude about life.

This segment is an appropriate supplement to your coverage of the nature-nurture issue, the relevance and ethics of animal testing, parental style, and other issues pertaining to the psychosocial development of children and the stability of personality throughout the life span.

Topics for Discussion

- psychology's "big issues" (nature-nurture, stability-change)
- research issues (adoption studies, animal research, ethical issues in research)
- psychosocial development (temperament, origins of personality, parental style

Discussion Questions

1. What two personality styles have been identified in rhesus monkeys? How do these styles relate to a monkey's relationship with his or her mother?
2. How did Steve Sumi and his team of researchers investigate the origins of these personality styles? What were their major findings?
3. What do you conclude from these studies regarding the stability of personality and the impact of parental style on the behavior of children?
4. Are animal studies useful in broadening the knowledge base of psychology and contributing to our understanding of human development?

Segment 24

Baby Body Sense (11:00)

Since Myrtle McGraw's pioneering experiments in the 1930s, psychologists have been fascinated by the question of how much of a baby's abilities are inborn and how much are the result of learning. This segment opens with archival footage of McGraw and several of her landmark experiments, including one in which a 2-week-old child is dunked into a pool to determine if the ability to swim is inborn.

In another longitudinal study, McGraw compared the motor development of identical twins, one of whom was introduced to roller skates at 13 months and given other special physical training. Although he outperformed his untrained twin on tests of physical strength and agility (on the basics of sitting, crawling, and walking), the special training did not accelerate the rate of development.

Indiana University psychologist Esther Thelan has spent years studying motor development in babies. We see footage of 3-month-old Madeleine, who doesn't quite know how to control her arms to grab a toy. Katherine, however, has no problem with the same skill. By recording muscle contractions and videotaping the babies from different angles, Thelan has found that, although all babies eventually master purposeful reaching, each reaches his or her goal in a unique way. This suggests to Thelan that reaching ability isn't simply programmed into genes.

The same is true for walking. "When a baby takes his or her first step," notes Thelan, "it looks as though the behavior just suddenly appears. But actually the baby has been working on that problem for a year."

As Thelan holds 7-month-old Eli above a treadmill, we see the stepping reflex, which many thought disappeared a few months after birth. Although stepping may be built in, babies' legs have to figure out how to stand and balance on their own. Observing footage of children of various ages, we see the gradual emergence of this complicated ability.

This segment is an excellent supplement to your coverage of motor development in children. It also provides a vivid example of how developmental psychologists have tackled the "big issue" of the relative importance of environmental and biological factors in development.

Topics for Discussion

- early studies of infant development
- nature-nurture controversy
- newborn reflexes
- sequence and timing of motor development

Discussion Questions

1. How did early developmental psychologists, such as Myrtle McGraw, investigate the relative importance of learning and heredity in the acquisition of physical skills? What did they find?
2. How has Esther Thelan approached the nature-nurture issue in her studies of directed reaching and walking? What has she found?

Segment 25

The Magic Years (10:00)

Magic is a perfect way to study the developing mind. At the University of Illinois, psychologist Carl Rosengren is investigating where preschoolers' ideas about magic originate. After analyzing videotapes of parents watching magic shows with their children, Rosengren has concluded that parents provide special explanations for extraordinary people and events. "Parents build up all of these stories about people who can do all sorts of things—tooth fairies, Santa Claus, magicians—people who have special powers that differentiate themselves from other individuals in our culture. Without the parents providing some sort of support for that, it's unlikely that the child will come up with these kinds of explanations entirely on their own."

But there's more to magic than just parental suggestion. Kids have to be ready to believe. One segment shows an experiment in which a complicated deception makes it appear to children that a magical machine has shrunk a room and its contents. Far from being amazed, 3-year-old Andrew explores the shrunken room as though nothing has happened. Rosengren explains: "One of the important things that children must learn is what kinds of things are possible in the world; until the child differentiates those things that are possible from those things that are not, there's no room for magic."

As an observer records 3-month-old Holly's gaze time, we see that she quickly loses interest in a nonmagical, real event. (A doll moves across her field of vision.) But when the doll magically seems to disappear and reappear, Holly stares much longer. She is surprised. Even at 3 months of age, she knows the world doesn't work this way.

However, Felix, who is just 2 weeks older than Holly, is not surprised. "These babies spontaneously came to the conclusion that we were using two different objects to produce the event. . . . It's absolutely remarkable that such little babies when shown surprising events are actively thinking about what we showed them and are actively searching for and finding explanations for what they see. It gives us a fascinating insight into what babies are doing when they look at the world around us."

If babies can be so logical, why do kids believe in magic? The answer seems to be that at around 5 or 6 there are still gaps in children's knowledge about the physical world. They are prepared to fill those gaps with a sort of catch-all explanation: "It must be magic." By age 7, a firm sense of reality has set in, and children are able to explain how a disappearing act is done—that it is a trick rather than magic.

This segment makes an interesting supplement to your discussion of cognitive development during childhood. It also presents a thorough description of the fascinating methodology and findings of recent studies of infant memory.

Topics for Discussion

- cognitive development in children
- infant memory
- methodology of developmental research

Discussion Questions

1. What have Carl Rosengren's studies of magic revealed about cognitive development in children? Why does the appreciation of magic depend on a person's age?
2. How has Renee Baillargeon studied infant cognitive development? In your answer, be sure to explain her research paradigm and outline her findings.

Segment 26

A Change of Mind (12:00)

Both developmental psychologists and parents know that preschoolers are quite limited in their ability to take into consideration another person's perspective. Preschoolers innocently believe that others see the world just as they do. These limitations are the subject of this video segment, which is hosted by Alan Alda.

At the University of Toronto, psychologists play a simple sorting game with several 3-year-old children. The object of the game is to see if the children can sort pictures based on certain criteria. In the first game, 3-year-old Jonathan has no trouble sorting boats and bunnies by their shape. Yet when he is asked to sort the same cards by their *color,* he persists in sorting by shape, as he did in the first game.

When 3-year-old Libby is asked to sort the cards first by color, she performs perfectly. But when she is asked to sort by shape, she gets stuck on the first rule she learned and continues to sort by color. The errors made by Jonathan and Libby demonstrate that it's the sequence of tasks that gives preschoolers trouble.

Another game played with 4½- and 5-year-olds reveals that they begin to think about other people's thoughts at this age. They learn that other people's thoughts vary, are private, and are sometimes incorrect.

This segment is an appropriate supplement to your coverage of cognitive development during the play years. It vividly illustrates the formation of mental sets, egocentric thinking, centered thinking, and the transition of preschool thought from limited perspective-taking to a theory of mind. The child's ability to deceive others and the appropriateness of various developmental tests at specific ages are also pertinent topics for discussion.

Topics for Discussion

- preschool egocentrism
- centered thinking
- the development of a theory of mind
- cognitive development and the ability to deceive others

Discussion Questions

1. In what ways is the mind of the typical 3-year-old unlike that of older children?
2. What characteristics of cognition during the play years were illustrated by the preschoolers as they played the card-sorting and doll games?
3. Why was 4½-year-old Patrick able to win when playing the "mean monkey game" whereas 3-year-old Jacob was not?
4. Were the various games appropriate tests of cognitive development in preschoolers? Did the nature of the games, or the instructors, limit or in any way influence the children's performance?

Segment 27

Teaching the Computer to Think (10:05)

Ask 7-year-old Anthony a common-sense question like, "Why can't a kid your age be a waiter in a restaurant?" and he'll have no trouble giving you a thoughtful answer. But ask him how many times he breathes in a day and he may wildly underestimate the correct answer. In contrast, a computer can easily be programmed to do the necessary math to answer the latter type of question, but it would have a lot more trouble answering the former question.

Would it be possible to program computers to develop a common-sense approach to the world? A team of programmers in Austin, Texas, thinks so. Project "cyc"—which is short for "encyclopedia"—has two objectives: (1) to program a computer with knowledge that people share in common about the world, and (2) to develop a machine that is capable of advancing its own learning.

So far, cyc's success has been limited. It can, however, engage in a simple form of logical reasoning. Cyc is not only learning common-sense knowledge but is also learning common-sense reasoning. If it is told, for example, that Karen is a musician and is married to John, it can infer that "John is probably a musician as well." It can do this because it knows the common-sense rule that spouses tend to have similar interests. This is a simple ability, but it demonstrates that cyc can figure things out, and, in a limited way, learn on its own.

At night, when cyc's programmers are sleeping, cyc is programmed to search for such connections among the bits of information stored in its knowledge base. Sometimes the connections can be quite creative, as when cyc came up with the analogy that the head of a family is like the dictator of a country.

This segment is appropriate as a supplement to your discussion of both cognitive development and artificial intelligence. It can also be used to illustrate several principles of adult thought processes.

Topics for Discussion

- cognitive development during childhood
- human problem solving
- heuristics and algorithms used in problem solving
- artificial intelligence

Discussion Questions

1. In what ways are human cognition and computer information-processing alike? In what ways are they different?
2. What is "cyc"? What are the objectives of this project? How are the project researchers carrying out their objectives?
3. Can a computer think? Explain your reasoning.

Segment 28

Superpower Ping-Pong (13:08)

In this segment, the winners of national contests from the United States, Japan, Germany, and Great Britain meet in Boston for the "World Series of Engineering." Each team has ten days to build a machine that collects a batch of Ping-Pong balls and somehow conveys them to a hopper. The team that gets the most balls in the hopper wins.

The students are divided into international teams. Contest organizer Harry West notes that "Today, and even more so in the future, engineering is going to be an international business. Products are designed by teams made up of engineers from all over the world. To be competitive, you have to be able to work together as a team."

The first task is to determine the team's goal: winning or creativity? Winning, however, is not really the point of the contest. It's about understanding other cultures, communicating ideas, and working together as part of a team. "These contests are spreading like wildfire," notes segment host Woody Flowers, "from elementary schools to universities all over the world. That's because they work. They teach physics, math, engineering, and other things like international cooperation. But most of all they link teaching to the real world."

This video segment is an appropriate supplement to your coverage of social and cultural issues in group behavior.

Topics for Discussion

- culture and learning
- cooperative learning
- group dynamics
- cooperation
- conformity
- superordinate goals

Discussion Questions

1. What are the objectives of the "World Series of Engineering"? In what ways do these objectives tie in with changes in the nature of engineering as a business?
2. Why are learning contests like this one catching on in schools? What advantages do they offer over more traditional methods of instruction?
3. Have you ever participated in a cooperative learning experiment? What was your experience like?

Segment 29

The Power of Persuasion (11:17)

As the segment opens, we observe a group of children at play. Some children have more influence over the group than others; they get what they want more often and seem to be natural leaders. Colgate University researchers are attempting to identify the origins of leadership by observing kids at play and tallying behaviors that indicate dominance.

Psychologist Carrie Keating has designed a fascinating experiment to test her theory that "in real-world leadership situations, it's frequently not the words that distinguish the leaders; they don't often have the best words, or the best ideas . . . What they have is a way to move us. What we're really studying is a little chunk of that charisma that defines leaders."

In Keating's experiment children are given a sip of sour juice that has been laced with salt and baking soda. They are then asked to pretend that they actually like the juice in order to persuade another person to try it.

As the videotape reveals, the nonverbal persuasion skills of submissive children are not very effective. They "leak" nonverbally with gestures and facial expressions—very subtle body movements such as scratching themselves or picking at their clothing. What happens when kids grow up? Does the connection between leadership and nonverbal persuasion remain?

In another experiment, male college students are placed in groups and given the task of figuring out how to survive a plane crash. It's not the ideas that are important to the researchers, it's who emerges as a leader.

Chris, who is a terrible liar, has some good ideas, but Keating says that's not what it takes to be a group leader. "That type of person may not be the person with the best ideas; it's the person who is the best at maneuvering and manipulating the group members, helping them along with their ideas so that they have moved toward some consensus."

Surprisingly, when Keating has conducted the experiment with women as the subjects the results have been entirely different. She has found no relationship between deception and leadership.

The segment ends with a chilling conclusion for this age in which the public has lost confidence in corporate and political leaders. Host Woody Flowers notes that Keating's research "is not just another knock on politicians; it's much more troubling. What it's telling us is that the skills that make people persuasive can also make us trust them even when they are lying."

This chapter touches on several issues that pertain to personality and social processes.

Topics for Discussion

- origins and stability of temperament
- social dominance
- leadership
- persuasion
- nonverbal communication
- deception

Discussion Questions

1. What has Carrie Keating's research revealed regarding the relationship between social dominance and deception in children?
2. Explain the gender difference that Keating has found in the relationship between leadership and deception. Can you think of a plausible explanation for why such a gender difference might have evolved? What possible advantages might it have conferred on to our species?
3. What are some of the practical implications of Keating's findings regarding leadership and deception? What warning does the segment host offer?

Segment 30

Virtual Fear (7:57)

For people who have a fear of heights, taking an elevator ride can be a nightmare. Systematic desensitization is an effective way to treat phobias. But in the twenty-first century, the computer may revolutionize traditional psychotherapy for phobias.

At Georgia Tech, Larry Hodges has built a virtual version of a hotel elevator. The helmet that host Alan Alda puts on his head gives him a computer-generated, three-dimensional view that moves when he turns his head. It's remarkably realistic. But is it realistic enough to help someone like Christopher Clock, who has had a debilitating fear of heights since he was a child and panicked when he tried to climb the Statue of Liberty?

Working with Emory University psychologist Barbara Rothbaum, Clock attempts to conquer his fear of heights. Rothbaum is collaborating with the Georgia Tech virtual-reality researchers to determine if patients with phobias find the virtual height as scary as the real thing. As Chris demonstrates, the answer is a definite "yes." This means that the technique can be used therapeutically. After eight sessions working with the elevator simulator, Chris has nearly conquered his fear. "I can actually look over the edge of the bridge and not be terrified." The first time Chris used the simulator he rated his fear at 100 (on a 100-point scale). Now he gives it a 25.

The real proof of the success of virtual therapy, however, comes when Chris is able to travel to the 72nd floor of an actual building in an external elevator, which bombards him with cues to height.

The Georgia researchers are now constructing a virtual airplane to help people overcome their fear of flying without having to take real flights with a therapist on board.

With the costs of computing power coming down fast, it won't be long before virtual-reality systems are cheap enough for every therapist to have one in his or her office.

This segment provides a captivating supplement to your discussion of psychotherapy, especially as it pertains to the treatment of phobias.

Topics for Discussion

- phobias
- systematic desensitization
- counterconditioning
- virtual reality and psychotherapy

Discussion Questions

1. What are phobias? What type of psychotherapy is most effective in treating this type of disorder?
2. How is the computer revolutionizing psychotherapy for phobias? Does the new technique have the same theoretical underpinnings as traditional psychotherapy for phobias? Explain your reasoning.
3. What are some other possible applications of virtual-reality technology to the treatment of psychological disorders?

Segment 31

Arachnophobia (9:31)

This segment opens with a day in the life of Joanne, an arachnophobe whose fear of spiders was once so extreme that it completely dominated her life, forcing her to wear heavy protective clothing, to tape openings around windows and doors obsessively, and to carry a special spider brush with her wherever she ventured.

But after watching a 1996 Scientific American Frontiers program on the use of virtual reality (VR) to help people conquer their fear of heights (see segment 31), Joanne asked her therapist if VR could be used to treat arachnophobia. Within a few weeks, University of Washington VR researcher Hunter Hoffman adapted an existing VR environment into one appropriate for systematically desensitizing arachnophobes.

Watching host Alan Alda flinch as he encounters a spider in a test of the virtual world are Joanne and her therapist, Al Carlin. After achieving an initial level of desensitization with the spider on a countertop, Alda moves to the second stage of VR therapy. Now, a larger spider hangs from a spider web directly in front of Alda's hand. Urged to touch the "spider" with his virtual hand, Hoffman places a realistic rubber spider in front of Alda, whose nervous laughter upon touching the "real" virtual insect, reveals an unexpectedly strong anxiety reaction.

After going through virtual systematic desensitization, Joanne is able to lead a normal life, free of her debilitating phobia, protective clothing, and compulsive behaviors. Brushing back tears of relief, she exclaims that she's even been able to go camping for the first time. As the ultimate test of her recovery, she is able to maintain her composure as her therapist places a real, live tarantula in her hand. Joanne's bravery inspires a reluctant Alda to hold the spider as a test of his own mastery over the fear of spiders.

This segment provides another captivating supplement to your discussion of psychotherapy and behavior modification, especially as they pertain to the treatment of phobias.

Topics for Discussion

- phobias
- systematic desensitization
- counterconditioning
- virtual therapy

Discussion Questions

1. What are phobias? How is systematic desensitization used to treat these anxiety disorders?
2. What is virtual reality? How is VR being used to revolutionize the treatment of phobias? What advantages does VR therapy offer over conventional therapy for phobias?

Segment 32

Cop Psychiatrists (10:30)

In New York City, when the police need help dealing with extraordinary circumstances they call the Emergency Service Unit. The officers in this unit can handle nearly any problem, from animal control to underwater recovery to working on top of sky scrapers.

But equipment is not always what is needed. In 1984, when the Emergency Service Unit was called to deal with a mentally ill woman, the woman was shot when she rushed an officer. Her death led to the realization that a new tool was needed. The tool that was created was a training course at John Jay College in which professional actors and psychologists guide the officers through the psychiatric disorders that they are likely to encounter on the job.

One actress displays the rage and delusions of bipolar disorder. Another portrays the intelligent and manipulative behavior of the classic psychopath. A third illustrates the telltale hallucinations and garbled speech of schizophrenia.

Joyce St. George leads the training. "It's going to help you get one more tool . . . if this person is delusional, if this person is talking to himself or herself, if its 95 degrees out and they've got a ski jacket on . . . These might be indicators that there's schizophrenia going on. If that's the case that might help you determine how to work with the person."

"Just the other day," notes Sergeant Mike Crowley, "we knocked out a peep hole and I looked in and the worst scenario was that we had a barricade who had offed his mother inside. But I could see that this guy had all the classic symptoms of schizophrenia . . . and I knew from the class that schizophrenics of this type are probably not violent, and I told the captain . . . and it encouraged the captain to allow us to pop the door and go in, and he went like a lamb."

The next scene presents two examples of the science of the classroom meeting the chaos of the street. On the first emergency call, Officer Crowley thinks the barricaded man is suffering from cocaine psychosis. Crowley tries to ease the situation, but this time violence erupts, and the man must be forcibly restrained.

On the second call, a severely disturbed man is holding his four children at gun point. "He's at a certain level of aggression," notes Officer Steve Green. "The last thing I want to do is raise that, then the children could be in definite peril. The class was very good at giving us a guideline to work by, what to listen for, what might set him off, what could steer him in the direction we wanted." Fortunately, on this call reasoning prevails, and the hostages are released unharmed.

This segment can be used to supplement your discussion of psychological disorders. It is also a good example of how psychology can be applied to real-world problems.

Topics for Discussion

- schizophrenia, bipolar disorder, cocaine psychosis
- applied psychology

Discussion Questions

1. What is the rationale for training police officers to differentiate among the symptoms of various mental disorders? Is the training working?
2. Can you think of other ways in which psychological research could be applied to police work?

Segment 33

Sports Imports (5:38)

The fast-growing field of sports science—including the subfield of sports psychology—is the subject of this segment, which is filmed at the training ground for Israel's top athletes.

Boris Blumenstein is coaching Alex, who represented Israel in the Olympic games. Boris has developed a biofeedback technique to help athletes like Alex improve their concentration. Electrodes placed on scalp muscles measure Alex's state of arousal. After a brief period of relaxation, Boris plays a videotape of Alex's last wrestling match. As Alex watches, his excitement level rises and his concentration falters.

The goal is control. Boris is teaching Alex to raise and lower his excitement level at will. This will give the athlete an important concentration tool that can be used during actual matches.

In the next segment sports scientists who specialize in biomechanics are helping a weight lifter who has reached a plateau in his training. Using sensors to analyze his motions, a computer constructs a model of the optimal weight-lifting motion. As they compare this with the weight lifter's actual motion, it is easy to see what the weight lifter is doing wrong—he's leaning too far forward, and his knees are not bent enough. Although it looks obvious on the computer, the problem was too subtle for the coaches to see during training.

The scientists are also working with a young sprinter. A sensor in his running shoes records his stride as he runs. The sensors reveal stride length, how much time between each step, when the heel and toe of each foot touch down, and so forth. The problem they are attempting to solve is the runner's inconsistent speed.

Once again, the computer compiles this information using the runner's height and weight and comes up with his ideal running tempo and a training regimen to improve his running form.

This segment provides a brief introduction to the new field of sports psychology. It can profitably be used as a supplement to your overview of psychology's subfields. It also touches on biofeedback and several issues pertaining to motivation and the management of arousal and stress.

Topics for Discussion

- sports psychology
- relaxation training
- optimal arousal theory
- biofeedback

Discussion Questions

1. How is the new field of sports psychology attempting to improve athletic performance? What specific examples are highlighted in the video segment?
2. What is the relationship between arousal and athletic performance? How are sports psychologists using biofeedback and relaxation training to help athletes manage their arousal?